梁燕 編著

新手入廚系列

魚味
無窮

前言

魚的品種非常多，簡單而言，可分為兩類：近海地區出產的，稱為海魚；內陸河流或魚塘出產的，稱為淡水魚。魚的營養豐富，不論家常便飯或宴會請客，都會以鮮魚作為材料，婚宴嫁娶、生日喜宴的酒席上，魚更是一道主菜。

正因為如此，魚有很多種做法，可分為蒸、炸、燜、焗、煎、炒、燒等，最後更返璞歸真，流行生吃，亦即刺身。在味道製作調配方面也十分豐富，可以配以甜、酸、苦、辣等味道，菜式多彩多姿，令人垂涎三尺。

近年，有很多戶外活動愛好者，每逢假日相約乘船出海，到達一些漁排養殖場垂釣，因為養殖場飼養着很多海魚，例如火點、石斑等，由於收穫比較豐富，感覺特別興奮，再交由漁家代為烹調，然後安坐漁排上，在充沛的陽光下，迎着海風，享用着最新鮮、最美味的海鮮，實在是一種難以形容的享受。

從健康的角度去衡量，魚含有豐富的蛋白質、維他命、鈣、碘及礦物質，脂肪含量少，能有效減低膽固醇，適合一家大小食用。

目錄

魚的選擇及處理方法

選擇

只要該魚的魚鱗發亮且緊黏魚身，魚鰓鮮紅，肉色潔白明亮，眼睛晶瑩，則可以放心購買。

處理

① 整條魚放冰箱，不用清洗。

② 魚洗淨後，用少許鹽塗抹均勻，全身包保鮮紙，放雪櫃中。

魚肉

① 製作魚肉時，宜選用肉質較結實的魚，例如鯪魚或鮫魚，肉質太散的魚如黃花魚就不能做魚膠。

② 魚劏洗淨，抹乾水分，去皮去骨，剁碎，放大碗中，加入調味料，順一個方向攪至起膠，期間分 3~4 次加入適量清水。

蒸魚

① 魚劏洗淨，抹乾水分，在碟上鋪上葱段，然後再放上魚，可令魚較快熟。

② 450 克魚大約蒸 10 分鐘（水沸才開始計時間）。

③ 蒸魚時待水大滾才放魚，蒸魚要用猛火才能鎖住肉汁，令魚肉更滑。

④ 蒸魚中途不要揭起鑊蓋。

⑤ 將蒸魚的魚汁即時倒去，再煮豉油汁和薑葱絲，淋在魚上。

煎魚

① 魚劏洗淨，以醃料醃好，抹乾水分，準備煎魚時才拍上生粉（可令魚更香脆和不會黏鑊）。

② 鑊要先燒熱至有煙，加入薑片爆香才放魚，然後轉慢火，耐心煎至金黃。

③ 留意魚眼轉白時，即可反轉另一面煎，大約煎 3~4 分鐘。煎魚不得其法，可能令魚皮黏鑊而破爛。

炸魚

① 魚劏洗淨，抹乾水分。

② 準備炸魚時才可黏上生粉或脆漿，因為生粉放在魚身上太久會黏在一起，難以炸脆。

③ 炸魚時，待油大滾才放魚，待魚定形後，轉中火炸至金黃即可盛起隔油；注意別炸至咖啡色，否則盛起後顏色會更深。

泰式煎魚

Pan-fried Fish in Thai Style

◯◯◯ 材料 | Ingredients

鮫魚 3 件
蒜頭 10 粒
葱 2 條
紅辣椒 2 隻
青檸 2 個

3 pieces mackerel
10 cloves garlic
2 stalks spring onion
2 red chilies
2 limes

⟨⟨⟩⟩ 醃料 | Marinade

酒 2 茶匙
鹽 1/2 茶匙
胡椒粉 1/2 茶匙

2 tsps wine
1/2 tsp salt
1/2 tsp ground white pepper

⟨⟨⟩⟩ 汁料 | Sauce

糖 4 湯匙
魚露 2 湯匙
黃薑粉 1/2 茶匙
小茴粉 1/4 茶匙

4 tbsps sugar
2 tbsps fish sauce
1/2 tsp turmeric powder
1/4 tsp fennel powder

⟨⟨⟩⟩ 做法 | Method

1. 鮫魚洗淨，瀝乾水分，用醃料醃 20 分鐘。
2. 燒熱鑊，下油將鮫魚煎至兩面金黃，上碟備用。
3. 蒜頭切茸，紅辣椒切圈，青檸榨汁。葱切段，鋪在鮫魚上。
4. 再燒熱油鑊，爆香蒜茸、紅辣椒，放入青檸汁和汁料，煮滾，淋在鮫魚上即成。

1. Rinse mackerel and drain. Marinate for 20 minutes.
2. Heat oil in a wok. Fry the fish until both sides are golden. Dish.
3. Chop garlic and cut red chilies into rings. Juice limes. Section spring onion and spread over the fish.
4. Heat a little oil in a wok. Stir-fry chopped garlic and red chilies until fragrant. Add lime juice and the sauce. Bring to the boil. Pour over the fish and serve.

入廚貼士 | Cooking Tips
可以改用鯧魚代替鮫魚。

魚茸煎餅

Pan-fried Dace Cakes with White Sesame

4 人
Serves 4

30 分鐘
30 Minutes

煎 Pan-fry

材料 | Ingredients

鯪魚肉 300 克
青檸 1 個
白芝麻 1 湯匙

300g dace paste
1 lime
1 tbsp white sesame

醃料 | Marinade

麵粉 4 湯匙
澄麵 2 湯匙
黑椒粉 1/2 茶匙
鹽 1/2 茶匙
水適量

4 tbsps flour
2 tbsps Tang flour
1/2 tsp ground black pepper
1/2 tsp salt
Some water

做法 | Method

1. 鯪魚肉加醃料，順一個方向拌勻成魚肉粉漿。
2. 白芝麻以白鑊炒香，備用。
3. 青檸榨汁，備用。
4. 燒熱油，分次放下魚肉粉漿，煎成魚餅形狀，至兩邊金黃色盛起。
5. 灑上炒香的白芝麻及青檸汁即成。

1. Add marinade into the dace paste and stir in one direction until sticky.
2. Stir-fry white sesame in a wok without oil until fragrant and set aside.
3. Juice lime.
4. Heat oil in a wok. Add the dace paste by several times. Fry into fish cakes and until both sides are golden brown. Dish.
5. Sprinkle over stir-fried white sesame and lime juice. Serve.

入廚貼士 | Cooking Tips

煎好的魚餅可切成條狀，加入其他材料如蔬菜或瓜類拌炒。

11

泰式魚餅

Thai Fish Cakes

◯◯◯ 材料 | Ingredients

鯪魚肉 500 克
雞蛋 1 隻
芫荽碎 2 茶匙
葱碎 1 茶匙
糖適量
鹽適量

500g dace paste
1 egg
2 tsps chopped coriander
1 tsp chopped spring onion
Pinch of sugar
Pinch of salt

⦿⦿ 調味料 | Seasonings

洋蔥 1/2 個	1/2 onion
香茅碎 1 湯匙	1 tbsp chopped lemongrass
蒜茸 2 茶匙	2 tsps grated garlic
蝦膏 1 1/2 茶匙	1 1/2 tsps shrimp paste
乾紅椒碎 1 茶匙	1 tsp chopped dried red chili
青檸皮茸 1 茶匙	1 tsp grated lime peel
紅椒粉 1 茶匙	1 tsp paprika
黃薑粉 1/2 茶匙	1/2 tsp turmeric powder
小茴香粉或五香粉 1/4 茶匙	1/4 tsp fennel powder or five-spice powder
油 1/2 湯匙	1/2 tbsp oil

⦿⦿ 做法 | Method

1. 鯪魚肉攪碎後，加入雞蛋、芫荽、葱、糖及鹽，順一個方向攪至起膠。
2. 調味料放入攪拌機打成糊狀，將 4 湯匙調味料加入魚肉中拌勻。
3. 將魚肉做成直徑約 7 厘米魚餅。
4. 燒熱鑊，將魚餅煎或炸成金黃色，瀝乾油分即成。

1. Mix egg, coriander, spring onion, sugar and salt with dace paste. Stir in one direction until sticky.
2. Blend seasoning in a blender until becomes paste. Mix 4 tbsps of seasoning with dace paste.
3. Shape the dace paste into fish cakes of about 7cm in diameter.
4. Heat oil in a wok. Fry the fish cakes or deep-fry them until golden brown. Drain and serve.

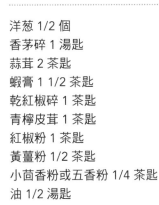

入廚貼士 | Cooking Tips

- 調味料要攪成爛糊，否則吃時會有硬塊。
- 500 克魚肉最多加 4 湯匙調味料，其顏色、味道及魚餅的質感最適合。

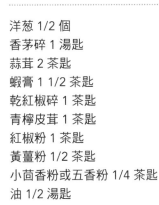

石斑文也

Grouper Menuiere

材料 | Ingredients

石斑肉 500 克
麵粉 250 克
雞蛋 5 隻

500g grouper flesh
250g flour
5 eggs

醃料 | Marinade

鹽 1 茶匙
胡椒粉 1/4 茶匙

1 tsp salt
1/4 tsp ground white pepper

4~6 人
Serves 4~6

30 分鐘
30 Minutes

汁料 | Sauce

牛油 100 克	100g butter
橙汁 200 毫升	200ml orange juice
忌廉 60 毫升	60ml whipping cream
檸檬汁 30 毫升	30ml lemon juice
糖 1 湯匙	1 tbsp sugar
麵粉 1 茶匙	1 tsp flour
鹽 1/4 茶匙	1/4 tsp salt

做法 | Method

1. 石斑肉洗淨，薄切雙飛或切 2 厘米厚件，加醃料拌勻醃片刻。

2. 雞蛋打勻，將石斑塊撲上麵粉，醮勻雞蛋液；燒熱油鑊，下石斑塊半煎炸至金黃色，隔油上碟。

3. 下牛油 1 茶匙於煲內煮熔，調入麵粉拌勻後，加入橙汁，以慢火略煮；再加入忌廉、糖，煮至汁液濃稠，調入檸檬汁和鹽拌勻，加入餘下牛油拌勻至熔，即可與魚塊同吃。

1. Rinse grouper flesh. Cut into slices and slit at the centre along the height without cutting through or cut into slices of 2cm thick. Marinate for a while.

2. Whisk eggs. Coat grouper with flour and then whisked eggs. Heat oil in a wok. Shallow-fry grouper until golden brown and drain. Dish.

3. Melt 1 tsp of butter in a pot. Add flour and mix well. Put in orange juice and cook over low heat briefly. Add cream and sugar. Cook until thickens. Mix in lemon juice and salt. Put in the remaining butter and mix until molten. Serve with the fish.

入廚貼士 | Cooking Tips

橙汁是用來調色的，可以改用其他的橙色材料。

4 人
Serves 4

30 分鐘
30 Minutes

酸辣多春魚

Pan-fried Spring Fish with Sour and Spicy Sauce

◯◯◯ 材料 | Ingredients

多春魚 400 克	400g spring fish
番茄 1 個	1 tomato
青椒 1/2 個	1/2 green bell pepper
葱 4 條	4 stalks spring onion
薑 3 片	3 slices ginger
蒜頭 1 粒	1 clove garlic
豆瓣醬 1 1/2 湯匙	1 1/2 tbsps spicy soybean sauce

◯◯◯ 調味料 | Seasonings

醋 3 湯匙	3 tbsps vinegar
老抽 2 茶匙	2 tsps dark soy sauce
酒 2 茶匙	2 tsps wine

⊘⊘⊘ 醃料 | Marinade

酒 2 茶匙
鹽 1 茶匙
胡椒粉 1 茶匙

2 tsps wine
1 tsp salt
1 tsp ground white pepper

⊘⊘⊘ 芡汁 | Thickening

生粉 1 茶匙
水 2 茶匙

1 tsp caltrop starch
2 tsps water

⊘⊘⊘ 做法 | Method

1. 所有材料（豆瓣醬除外）洗淨，番茄切粒，青椒去籽切粒，蒜頭切茸，葱切粒，薑一半切粒，一半切片。
2. 多春魚洗淨，抹乾水分，加醃料醃 30 分鐘，撲上生粉。
3. 燒熱鑊，下油爆香薑片，將多春魚煎至兩面金黃，盛起。
4. 下油爆蒜茸、薑粒、豆瓣醬、番茄、青椒，再加入魚及調味料煮片刻，勾芡，下葱兜勻，即可上碟。

1. Rinse all ingredients. Dice tomato. Seed green bell pepper and dice. Grate garlic and dice spring onion. Dice half amount of ginger and slice the other half.
2. Rinse spring fish and wipe dry. Marinate for 30 minutes and coat with caltrop starch.
3. Heat oil in a wok. Stir-fry ginger slices until fragrant. Add spring fish and fry until both sides are golden brown. Dish.
4. Heat a little oil in a wok. Stir-fry grated garlic, ginger dice, spicy soybean sauce, tomato and green bell pepper for a while. Add fish and seasoning. Cook for a while. Thicken the sauce with caltrop starch solution. Add spring onion and mix well. Serve.

入廚貼士 | Cooking Tips

多春魚可原條煎，不用取出腸臟。

涼瓜煎鱠魚

Pan-fried Pomfret with Bitter Melon

(CO) 材料 | Ingredients

鱠魚 1 條
涼瓜 240 克
水適量

1 pomfret
240g bitter melon
Some water

4~6 人
Serves 4~6

20 分鐘
20 Minutes

醃料 | Marinade

酒 1/2 茶匙
胡椒粉 1/2 茶匙
鹽 1/4 茶匙

1/2 tsp wine
1/2 tsp ground white pepper
1/4 tsp salt

調味料 | Seasonings

蒜茸 1 湯匙
豆豉 2 茶匙
鹽 1/2 茶匙
糖 1/4 茶匙

1 tbsp grated garlic
2 tsps fermented black beans
1/2 tsp salt
1/4 tsp sugar

做法 | Method

1. 鱠魚劏洗淨，抹乾水分，用醃料醃 10 分鐘，魚面剶 2 刀。
2. 燒熱鑊，下鱠魚煎至兩面金黃，瀝乾油分，盛起。
3. 涼瓜洗淨，去瓢去核，切成骨排形。
4. 再燒熱油鑊，爆香調味料，加入涼瓜、水煮片刻，慢火燜至涼瓜軟身，將鱠魚回鑊，用中火燜至水分收乾，上碟即可。

1. Gut pomfret and wipe dry. Marinate for 10 minutes and slit twice on the surface.
2. Heat oil in a wok. Fry pomfret until both sides are golden and drain. Dish.
3. Rinse bitter melon. Remove pith and core. Cut into small rectangular pieces.
4. Heat a little oil in a wok. Stir-fry seasoning until fragrant. Add bitter melon and water. Cook for a while. Simmer over low heat until the melon is soft. Put back pomfret and simmer over medium heat until dry. Serve.

入廚貼士 | Cooking Tips

涼瓜除了去瓢去核外，更要去除白色的部分才不苦。

4~6 人
Serves 4~6

30 分鐘
30 Minutes

狗肚魚煎蛋

Pan-fried Bombay Duck
Fish and Eggs

材料 | Ingredients

狗肚魚 6 條
雞蛋 4 隻

6 Bombay duck fish
4 eggs

醃料 | Seasonings

酒 1/2 茶匙
胡椒粉 1/2 茶匙
鹽 1/2 茶匙

1/2 tsp wine
1/2 tsp ground white pepper
1/2 tsp salt

做法 | Method

1. 狗肚魚洗淨，切去頭尾，取去中骨起肉，抹乾水分，用醃料醃好。
2. 燒熱油鑊，下狗肚魚煎至金黃，盛起，瀝乾油分，備用。
3. 雞蛋打勻，放油中煎至半熟，加入狗肚魚拌勻，煎至金黃即可。

1. Rinse Bombay duck fish and cut off heads and tails. Bone and wipe dry. Marinate for a while.
2. Heat oil in a wok. Fry the fish until golden brown and drain. Dish.
3. Whisk eggs. Fry in oil until medium-cooked. Add Bombay duck fish and mix well. Fry until golden brown and serve.

入廚貼士 | Cooking Tips

狗肚魚用醃料醃好也可放油中炸。

炒鱔糊

Stir-fried Eel with Mungbean Sprouts

⚬⚬⚬ 材料 | Ingredients

黃鱔 500 克
銀芽 150 克
蒜茸 4 粒量
薑茸 1 茶匙
酒少許
胡椒粉少許

500g yellow eel
150g mungbean sprouts
4 cloves garlic (grated)
1 tsp grated ginger
Some wine
Pinch of ground white pepper

4~6 人
Serves 4~6

20 分鐘
20 Minutes

⊙⊙⊙ 調味料 | Seasonings

老抽 2 湯匙	2 tbsps dark soy sauce
麻油 1 茶匙	1 tsp sesame oil
生粉 1 茶匙	1 tsp caltrop starch
糖 3/4 茶匙	3/4 tsp sugar
鹽 1/2 茶匙	1/2 tsp salt
水 1/4 杯	1/4 cup water

⊙⊙⊙ 做法 | Method

1. 黃鱔洗淨，起骨去頭尾，汆水去黏液，再洗淨後切成約 4 厘米長的條狀。
2. 銀芽洗淨，瀝乾。燒熱鑊，下油炒熟銀芽，盛起備用。
3. 下油爆香蒜茸、薑茸，倒入黃鱔炒香，灒酒，加入調味料，倒入銀芽炒勻後上碟，加入少許胡椒粉拌勻，即可。

1. Rinse yellow eel. Bone and remove head and tail. Scald for a while to remove the slime. Rinse and cut into strips of about 4 cm long.
2. Rinse mungbean sprouts and drain. Heat oil in a wok and stir-fry mungbean sprouts until done. Dish.
3. Heat oil in a wok. Stir-fry grated garlic and grated ginger until fragrant. Pour in yellow eel and stir-fry until fragrant. Sizzle in wine. Add seasoning. Put in mungbean sprouts and stir-fry until well-mixed. Mix in a little ground white pepper and serve.

入廚貼士 | Cooking Tips

可將蒜茸另外炸至金黃，留待上碟時灑上便更香脆。

鮮蝦炒鱔片

Stir-fried Eel with Fresh Prawns

◯◯◯ 材料 | Ingredients

黃鱔 400 克	400g yellow eel
大蝦仁 150 克	150g large shelled prawns
蒜頭 3 粒	3 cloves garlic
酒少許	Some wine

◯◯◯ 醃料 | Marinade

胡椒粉少許
生粉少許
鹽少許

Pinch of ground white pepper
Pinch of caltrop starch
Pinch of salt

⭕⭕⭕ 調味料 │ Seasonings

老抽 1 1/2 茶匙	1 1/2 tsps dark soy sauce
生抽 1 茶匙	1 tsp light soy sauce
麻油 1 茶匙	1 tsp sesame oil
糖 1/2 茶匙	1/2 tsp sugar
水 2 湯匙	2 tbsps water

⭕⭕⭕ 芡汁 │ Thickening

生粉水少許

Some caltrop starch solution

入廚貼士 | Cooking Tips

黃鱔可請魚販代為起骨及去潺。

⭕⭕⭕ 做法 │ Method

1. 黃鱔洗淨，起骨去頭尾，汆水去黏液，洗淨後切片，放入胡椒粉、生粉拌勻，備用。
2. 大蝦仁洗淨，抹乾水分，加入胡椒粉、鹽、生粉拌勻。
3. 燒熱鑊，下蝦仁泡油後盛起，備用。
4. 下油爆香蒜頭，放入鱔片炒熟，潵酒，倒入調味料，用少許生粉水勾芡，加入蝦仁炒勻即成。

1. Rinse yellow eel. Bone and remove head and tail. Scald for a while to remove the slime. Rinse and slice again. Mix in ground white pepper and caltrop starch. Set aside.
2. Rinse large shelled prawns and wipe dry. Mix in ground white pepper, salt and caltrop starch.
3. Heat oil in a wok. Jiggle shelled shrimps in oil slowly and drain.
4. Stir-fry garlic in oil until fragrant. Put in eel slices and stir-fry until done. Sizzle in wine and seasoning. Thicken the sauce with a little caltrop starch solution. Add shelled prawns and stir-fry well. Serve.

炒桂花魚肚

Stir-fried Dried Fish Maw with Mungbean Sprouts and Egg Whites

⬤⬤ 材料 | Ingredients

魚肚 100 克	100g dried fish maw
韭黃 100 克	100g yellow chives
銀芽 100 克	100g mungbean sprouts
蟹肉 100 克	100g crabmeat
蛋白 10 隻	10 egg whites
火腿絲 3 片	3 slices ham (shredded)
葱白 2 條	2 stalks spring onion (white part)
薑絲 2 片量	2 slices ginger (shredded)

⬤⬤ 煨料 | Condiments

薑 2 片	2 slices ginger
葱 2 條	2 stalks spring onion
酒 1 湯匙	1 tbsp wine

6 人
Serves 6

20 分鐘
20 Minutes

◯◯◯ 雞蛋調味料 | Seasoning for eggs

生粉 1 湯匙　　　　1 tbsp caltrop starch
薑汁酒 1 茶匙　　　1 tsp ginger wine
鹽 1 茶匙　　　　　1 tsp salt
糖 1/2 茶匙　　　　1/2 tsp sugar
胡椒粉 1/2 茶匙　　1/2 tsp ground white pepper
麻油 1/4 茶匙　　　1/4 tsp sesame oil

入廚貼士 | Cooking Tips
爆香的材料要放涼後才盛入蛋白內。

◯◯◯ 做法 | Method

1. 韭黃洗淨，切段。銀芽洗淨，瀝乾。
2. 魚肚洗淨，以清水浸透至軟身，用煨料煮片刻，以清水洗淨，切長條。
3. 燒熱鑊，下油爆香薑絲，加入銀芽拌炒，下鹽調味，盛起，備用。
4. 蛋白放大碗中，加入調味料拌勻，加入銀芽、火腿和蟹肉拌勻，備用。
5. 燒熱油鑊，爆香葱白，加入韭黃、魚肚，倒入蛋白混合物，快手兜勻，上碟即成。

1. Rinse yellow chives and section. Rinse mungbean sprouts and drain.
2. Rinse dried fish maw and soak in water thoroughly until soft. Cook together with condiments for a while. Rinse and cut into long strips.
3. Heat oil in a wok. Stir-fry ginger shreds until fragrant. Add mungbean sprouts and stir-fry well. Season with salt and dish.
4. Put egg whites into a large bowl. Put in seasoning and mix well. Add mungbean sprouts, ham and crabmeat. Mix and set aside.
5. Heat oil in a wok. Stir-fry white part of spring onion until fragrant. Add yellow chives and dried fish maw. Pour in the egg white mixture. Fry quickly and serve.

韭黃魚塊

Stir-fried Grouper Fillet with Yellow Chives

材料 | Ingredients

石斑肉 400 克
韭黃 100 克
葱 3 條
蘆筍 2 條
甘筍 1/2 條
薑 2 片

400g grouper fillet
100g yellow chives
3 stalks spring onion
2 asparaguses
1/2 carrot
2 slices ginger

炒
Stir-fry

⟨⟨⟨⟩⟩⟩ 醃料 | Marinade

蛋白 1/2 隻
酒 1 茶匙
鹽 1/2 茶匙
糖 1/2 茶匙
生粉 1/2 湯匙

1/2 egg white
1 tsp wine
1/2 tsp salt
1/2 tsp sugar
1/2 tbsp caltrop starch

⟨⟨⟨⟩⟩⟩ 芡汁 | Thickening

蠔油 1 湯匙
麻油 1 茶匙
胡椒粉 1 茶匙
生粉 1 茶匙
水 30 毫升

1 tbsp oyster sauce
1 tsp sesame oil
1 tsp ground white pepper
1 tsp caltrop starch
30ml water

⟨⟨⟨⟩⟩⟩ 做法 | Method

1. 石斑肉洗淨，切成約 2 厘米厚方塊，加醃料醃 30 分鐘。
2. 其他材料分別洗淨，葱、蘆筍、韭黃切段，甘筍切片。
3. 燒熱鑊，下石斑塊泡嫩油，撈起，瀝乾油分，備用。
4. 再燒熱油鑊，爆香葱、薑、甘筍、蘆筍等，炒香後倒入芡汁煮滾，加入韭黃，將魚塊回鑊，即可上碟。

1. Rinse grouper fillet and cut into pieces of 2cm thick. Marinate for 30 minutes.
2. Rinse other ingredients respectively. Section spring onion, asparaguses and yellow chives. Slice carrot.
3. Heat oil in a wok and jiggle grouper in warm oil slowly. Drain and set aside.
4. Heat a little oil in a wok. Stir-fry spring onion, ginger, carrot and asparaguses until fragrant. Pour in the sauce and bring to the boil. Add yellow chives and fish. Mix well and serve.

入廚貼士 | Cooking Tips
韭黃很快便煮熟，所以不可太早加入鑊中炒。

白汁吞拿魚意粉

Tuna and Spaghetti in White Sauce

4 人
Serves 4

20 分鐘
20 Minutes

⬤⬤ 材料 | Ingredients

意大利粉 250 克
罐裝吞拿魚 200 克（水浸）
忌廉湯 200 克
牛油 50 克
洋葱 1 個
蒜頭 5 粒

250g spaghetti
200g canned tuna in water
200g cream soup
50g butter
1 onion
5 cloves garlic

⬤⬤ 調味料 | Seasonings

鹽 1/2 茶匙
胡椒粉 1/2 茶匙

1/2 tsp salt
1/2 tsp ground white pepper

⬤⬤ 入廚貼士 | Cooking Tips

意大利粉不用煮得太熟，
因為離火後還有熱力可繼
續煮熟。

⬤⬤ 做法 | Method

1. 燒滾一鍋水，下少許鹽，將意大利粉放滾水中煮熟，盛起，瀝乾
   水分。
2. 蒜頭切片，洋葱去衣切粒。
3. 燒熱鑊，以慢火煮熔牛油，爆香蒜片、洋葱，加入吞拿魚、忌廉
   湯及調味料，煮滾後加入意粉炒勻，即可上碟。

1. Bring a pot of water to the boil and add a little salt. Cook
   spaghetti in boiling water until done and drain.
2. Slice garlic. Skin and dice onion.
3. Heat wok and melt butter over low heat. Stir-fry garlic slices and
   onion until fragrant. Add tuna, cream soup and seasoning. Bring
   to the boil. Put in the spaghetti and mix well. Serve.

黃花魚羹

Yellow Croaker Chowder

◯◯ 材料 | Ingredients

黃花魚 600 克　　　　葱 2 條
冬菇 4 朵（浸軟）　　薑 2 片
雞蛋 2 隻　　　　　　水 1,500 毫升
火腿 1 片

600g yellow croaker
4 dried black mushrooms
(soaked until soft)
2 eggs
1 slice ham
2 stalks spring onion
2 slices ginger
1,500ml water

醃料 | Marinade

酒 1 湯匙
胡椒粉 1/2 茶匙

1 tbsp wine
1/2 tsp ground white pepper

調味料 | Seasonings

胡椒粉 1 茶匙
鹽 1 茶匙
麻油 1 茶匙

1 tsp ground white pepper
1 tsp salt
1 tsp sesame oil

芡汁 | Thickening

生粉 3 湯匙
水 50 毫升

3 tbsps caltrop starch
50ml water

入廚貼士 | Cooking Tips

黃花魚可改用其他魚類代替，
但需要選肉質比較幼嫩的。

做法 | Method

1. 黃花魚洗淨，抹乾水分，用醃料塗勻魚肚魚身，隔水蒸 15 分鐘，
   去骨起肉。
2. 冬菇、火腿、葱洗淨，切細粒。
3. 燒滾水，下冬菇、薑片、魚肉、火腿和調味料煮片刻，勾芡，加
   入雞蛋拌勻，最後下葱粒，即可。

1. Rinse yellow croaker and wipe dry. Rub the whole fish including
   its stomach with the marinade. Steam for 15 minutes and bone.
2. Rinse dried black mushrooms, ham and spring onion. Dice them
   finely.
3. Bring water to the boil. Add dried black mushrooms, ginger
   slices, fish, ham and the seasoning. Cook for a while. Put in the
   caltrop starch solution and mix in eggs. Add diced spring onion
   and serve.

忌廉三文魚湯

Cream of Salmon Soup

4 人
Serves 4

40 分鐘
40 Minutes

材料 | Ingredients

三文魚 250 克	250g salmon
西芹 3 條	3 stalks celery
甘筍 1 條	1 carrot
洋葱 2 個	2 onions
麵粉 1 湯匙	1 tbsp flour
牛油 1 湯匙	1 tbsp butter
清雞湯 800 毫升	800ml clear chicken broth
鮮奶 500 毫升	500ml fresh milk
忌廉 120 毫升	120ml whipping cream
鹽適量	Pinch of salt

做法 | Method

1. 甘筍去皮切塊,洋葱去衣切塊,西芹、三文魚切小粒。
2. 牛油放煲內以慢火煮熔,下洋葱、西芹、甘筍,炒至軟身,拌入麵粉,慢慢加入鮮奶和清雞湯,以慢火煮滾,熄火待涼。
3. 將湯放入攪拌機內打成茸汁,隔渣後放回煲內,下鹽調味,再煮滾,拌入忌廉和三文魚肉粒,灼至僅熟,即可。

1. Peel carrot and cut into pieces. Skin onion and cut into pieces. Dice celery and salmon finely.
2. Cook butter over low heat in a pot until molten. Put in onion, celery and carrot. Fry until soft. Mix in flour and add in fresh milk and clear chicken broth slowly. Bring to the boil over low heat. Remove from heat and set aside to let cool.
3. Pour the soup into a blender and blend into juice like. Sieve out the condiments and put the soup back into the pot. Season with salt and bring to the boil again. Mix in cream and diced salmon. Cook until the salmon is just done. Serve.

入廚貼士 | Cooking Tips

飲湯時可加入炸脆的麵包粒或餅碎伴吃。

周打魚湯

Fish Chowder

材料 | Ingredients

石斑肉 300 克	300g grouper fillet
煙肉 4 片	4 slices bacon
洋葱 1 個	1 onion
西芹 1 棵	1 stalk celery
薯仔 1 個	1 potato
甘筍 1/2 條	1/2 carrot
鮮奶或忌廉 240 毫升	240ml fresh milk or cream
雞湯或清水 2,000 毫升	2 liters chicken broth or water
麵粉 6 湯匙	6 tbsps flour
牛油 30 克	30g butter
鹽、胡椒粉各適量	Pinch of salt
	Pinch of ground white pepper

Boil

香料 | Spices

香葉 1 片　　　　　1 bay leaf
千里香 1/2 茶匙　　1/2 tsp thyme

做法 | Method

1. 煙肉切碎，洋葱、甘筍、西芹、薯仔洗淨去皮切細粒。
2. 先將一半煙肉放油鑊中炸脆，備用。
3. 燒熱鑊，加入牛油，以慢火炒洋葱、甘筍、西芹，炒至金黃色，下麵粉拌勻，離火冷卻一會兒。
4. 將鑊放回爐上，慢慢加入雞湯或清水拌勻，加入薯仔、煙肉和香料，以中火煮 20 分鐘。
5. 徐徐加入鮮奶、鹽、胡椒粉，煮滾後灑下炸脆的煙肉即可。

1. Chop bacon. Rinse onion, carrot, celery and potato. Peel them and cut into fine dice.
2. Deep-fry half amount of bacon until crispy and set aside.
3. Heat butter in a wok. Stir-fry onion, carrot and celery over low heat until golden brown. Mix in flour. Remove from heat and set aside to let cool for a while.
4. Heat the wok again and add in chicken broth or water slowly. Mix well. Put in potato, bacon and spices. Cook over medium heat for 20 minutes.
5. Add in fresh milk, salt and ground white pepper slowly. Bring to the boil and sprinkle over the deep-fried bacon. Serve.

入廚貼士 | Cooking Tips
麵粉炒勻後，加入雞湯，用鑊鏟推開麵粉，直至麵粉全部溶在湯內。

椒鹽狗肚魚

Salt and Pepper Bombay Duck Fish

材料 | Ingredients

狗肚魚 500 克
葱粒 4 條量
紅辣椒 1 隻
蛋白 1 隻
麵粉 1 湯匙
淮鹽 1 茶匙
酒適量

500g Bombay duck fish
4 stalks spring onion (diced)
1 red chili
1 egg white
1 tbsp flour
1 tsp Huai salt
Some wine

4 人
Serves 4

30 分鐘
30 Minutes

◎◎ 醃料｜Marinade

酒 2 茶匙
胡椒粉 1 茶匙
鹽 1/2 茶匙

2 tsps wine
1 tsp ground white pepper
1/2 tsp salt

◎◎ 做法｜Method

1. 狗肚魚洗淨，去骨起肉，用醃料醃 2 小時。
2. 燒熱鑊，狗肚魚抹乾水分，加蛋白，撲上麵粉，放油中炸至金黃色。
3. 再燒熱鑊，下少許油爆紅椒，下一半葱，將狗肚魚回鑊，輕手兜勻，灒酒，灑下淮鹽和餘下的葱粒即成。

1. Rinse Bombay duck fish. Bone and marinate the flesh for 2 hours.
2. Heat oil in a wok. Wipe dry the fish and mix with egg white. Coat with flour and deep-fry the fish in oil until golden brown.
3. Heat a little oil in a wok. Stir-fry red chili until fragrant. Add half amount of spring onion and put in the fish. Fry lightly. Sizzle in wine. Sprinkle over Huai salt and the remaining spring onion dice. Serve.

入廚貼士｜Cooking Tips
狗肚魚要完全抹乾水分才下油鑊炸，才會香脆。

椒鹽脆肉鯇

Deep-fried Grass Carp Belly with
Salt and Chili

⟨⟨⟩⟩ 材料 | Ingredients

脆肉鯇魚腩 300 克
蒜頭 3 粒
紅辣椒 1 隻
淮鹽約 1 茶匙

300g grass carp belly (crispy)
3 cloves garlic
1 red chili
Approx. 1 tsp Huai salt

⟨⟨⟩⟩ 醃料 | Marinade

酒 1 茶匙
鹽 1/2 茶匙
胡椒粉 1/2 茶匙

1 tsp wine
1/2 tsp salt
1/2 tsp ground white pepper

⟨⟨⟩⟩ 做法 | Method

1. 魚腩洗淨，切成 3 厘米寬的長條形，抹乾水分，加入醃料醃 10 分鐘。
2. 蒜頭洗淨，剁茸；紅辣椒洗淨，切幼粒。
3. 燒熱油鑊，將魚腩炸至金黃色，盛起，瀝乾油分。
4. 再燒熱油鑊，爆香蒜茸、紅辣椒，將魚腩回鑊，再灑下淮鹽拌勻，即可上碟。

1. Rinse fish belly and cut into long strips of 3cm wide. Wipe dry and marinate for 10 minutes.
2. Rinse garlic and chop finely. Rinse red chili and dice finely.
3. Heat oil in a wok. Deep-fry the fish belly until golden brown and drain.
4. Heat a little oil in a wok. Stir-fry chopped garlic and red chili until fragrant. Put in the fish belly and sprinkle over Huai salt. Mix well and serve.

入廚貼士 | Cooking Tips
如怕味道太辣，紅辣椒不要爆香，最後才灑上，便有鮮艷顏色而沒有辣味。

酸甜荔枝魚腩

Deep-fried Fish Belly with Lychees and Sweet and Sour Sauce

材料 | Ingredients

魚腩 300 克
荔枝 8 粒
青、紅椒各 1 個（切塊）
菠蘿 3 圈

薑片 3 片
葱 3 條
蒜茸 2 湯匙
生粉、酒適量

300g fish belly
8 lychees
1 green pepper (cut into pieces)
1 red pepper (cut into pieces)
3 rings pineapple
3 slices ginger
3 stalks spring onion
2 tbsps grated garlic
Pinch of caltrop starch and some wine

6 人
Serves 6

30 分鐘
30 Minutes

◯◯◯ 醃料 | Marinade

酒 2 茶匙
胡椒粉 1 茶匙
鹽 1/2 茶匙

2 tsps wine
1 tsp ground white pepper
1/2 tsp salt

◯◯◯ 調味料 | Seasonings

白醋 2 湯匙
糖 1 湯匙
薑汁 2 茶匙
生粉 1 茶匙
鹽 1/2 茶匙
水 2 湯匙

2 tbsps white vinegar
1 tbsp sugar
2 tsps ginger juice
1 tsp caltrop starch
1/2 tsp salt
2 tbsps water

入廚貼士 | Cooking Tips
熄火後才將水果落鑊炒勻。

◯◯◯ 做法 | Method

1. 魚腩洗淨,切約 4 厘米寬的長塊,加醃料醃約 15 分鐘。

2. 燒熱鑊,魚腩撲上生粉,放入油鑊中炸至金黃色,瀝乾油分,備用。

3. 再燒熱油鑊,爆香蒜茸、薑片、青椒、紅椒,灒酒,下調味料煮成濃汁,將魚腩回鑊,熄火,下荔枝、菠蘿、葱兜勻,上碟。

1. Rinse fish belly and cut into long pieces of about 4cm wide. Marinate for about 15 minutes.

2. Heat oil in a wok. Coat fish belly with caltrop starch and deep-fry until golden brown. Drain.

3. Heat a little oil in a wok. Stir-fry grated garlic, ginger slices, green and red peppers until fragrant. Sizzle in wine. Add seasoning and cook into thick sauce. Put in the fish belly and remove from heat. Add lychees, pineapple and spring onion. Mix and serve.

福州香酥魚

Deep-fried Black Silver Carp

◯◯◯ 材料 | Ingredients

大魚 300 克

300g black silver carp

◯◯◯ 醃料 | Marinade

酒 1 茶匙
鹽 1/2 茶匙
胡椒粉 1/2 茶匙

1 tsp wine
1/2 tsp salt
1/2 tsp ground white pepper

◯◯◯ 炸漿 | Deep-frying paste

雞蛋 1 隻
生粉 1/2 杯
水約 2 1/2 湯匙

1 egg
1/2 cup caltrop starch
Approx. 2 1/2 tbsps water

入廚貼士 | Cooking Tips

炸漿攪勻後，在其上蓋一層
薄薄的油，可令漿發得更好。

◯◯◯ 做法 | Method

1. 大魚洗淨，抹乾水分，切成大約長 6 厘米、寬 3 厘米的方塊，加
   醃料醃 15 分鐘。
2. 炸漿攪勻成稀糊。
3. 燒熱油鑊，將大魚沾滿炸漿，放油中炸至金黃色，瀝乾油分即可
   上碟，食時可蘸沙律醬吃。

1. Rinse and wipe dry black silver carp. Cut into pieces of about 6
   cm long and 3cm wide. Marinate for 15 minutes.
2. Mix the deep-frying paste into dilute paste.
3. Heat oil in a wok. Coat the fish with deep-frying paste. Deep-fry
   it until golden brown and drain. Serve with salad dressing.

醬爆鰻魚

Braised Eel with Fermented Soybean Sauce

白鱔 600 克
乾葱頭 2 粒
磨豉醬 1 湯匙
薑粒 2 茶匙

600g white eel
2 shallots
1 tbsp fermented soybean paste
2 tsps chopped ginger

6 人
Serves 6

25 分鐘
25 Minutes

46

◯◯◯ 醃料 | Marinade

酒 1 湯匙
蛋白 1 湯匙
薑汁 1 茶匙
胡椒粉 1/2 茶匙
鹽 1/4 茶匙

1 tbsp wine
1 tbsp egg white
1 tsp ginger juice
1/2 tsp ground white pepper
1/4 tsp salt

◯◯◯ 調味料 | Seasonings

生抽 1 湯匙
糖 4 茶匙
胡椒粉 1/2 茶匙
水 6 湯匙

1 tbsp light soy sauce
4 tsps sugar
1/2 tsp ground white pepper
6 tbsps water

◯◯◯ 做法 | Method

1. 白鱔洗淨，去骨，汆水去黏液，抹乾水分，剝十字花，用醃料醃 15 分鐘。

2. 燒熱鑊，白鱔撲上生粉，放滾油中炸至金黃色，瀝乾油分，盛起。

3. 再燒熱鑊，爆香乾葱、薑粒及磨豉醬，下調味料，將鱔回鑊拌勻，即可上碟。

1. Rinse white eel and bone. Scald for a while to remove the slime. Wipe dry and slit crosses on the surface. Marinate for 15 minutes.
2. Heat oil in a wok. Coat eel with caltrop starch and deep-fry in boiling oil until golden brown. Drain. Dish.
3. Heat a little oil in a wok. Stir-fry shallots, chopped ginger and fermented soybean paste until fragrant. Add seasoning and eel. Stir-fry well and serve.

入廚貼士 | Cooking Tips

魚肉上面斜成十字花狀，炸後會捲起來。

西檸魚柳

Deep-fried Sole Fillet with Lemon Sauce

◯◯◯ 材料 | Ingredients

龍脷柳 2 條
檸檬皮茸 1 個量
牛油 1 湯匙
麵粉適量

2 pieces sole fillet
1 ground lemon peel
1 tbsp butter
Some flour

⊙⊙ 醃料 | Marinade

吉士粉 2 茶匙
生粉 2 茶匙
鹽 1 茶匙
糖 1 茶匙
酒 1 茶匙

2 tsps custard powder
2 tsps caltrop starch
1 tsp salt
1 tsp sugar
1 tsp wine

⊙⊙ 調味料 | Seasonings

檸檬汁 1 1/2 湯匙
鹽 1/2 茶匙
糖 1/2 茶匙

1 1/2 tbsps lemon juice
1/2 tsp salt
1/2 tsp sugar

⊙⊙ 做法 | Method

1. 龍脷柳洗淨，抹乾水分，切成塊狀，用醃料醃約 30 分鐘。
2. 燒熱油，將龍脷柳沾上麵粉，放油中炸至金黃色，盛起，瀝乾油分。
3. 再燒熱油，下牛油用慢火煮熔，加入檸檬皮茸和調味料煮勻。
4. 將龍脷柳回鑊，拌勻至沾上汁液，即可上碟。

1. Rinse sole fillet and wipe dry. Cut into pieces and marinate for about 30 minutes.
2. Heat oil in a wok. Coat fish fillet with flour and deep-fry in oil until golden brown. Drain.
3. Heat a little oil in a wok. Add butter and melt over low heat. Put in lemon peel and the seasoning. Cook until well-mixed.
4. Put sole fillet into the wok. Mix until coated with sauce. Serve.

入廚貼士 | Cooking Tips

準備炸魚柳時才沾麵粉，不可太早。

芝士斑塊

Deep-fried Grouper with Cheese Sauce

6 人
Serves 6

30 分鐘
30 Minutes

⊙⊙⊙ 材料 | Ingredients

急凍石斑肉 500 克
忌廉雞湯 150 克
芝士 4 片
生粉 3 湯匙
牛油適量

500g frozen grouper fillets
150g cream of chicken soup
4 slices cheese
3 tbsps caltrop starch
Some butter

⊙⊙⊙ 醃料 | Marinade

雞蛋 2 隻
鹽 1 茶匙
胡椒粉 1 茶匙
薑汁 1 茶匙
麻油 1 茶匙

2 eggs
1 tsp salt
1 tsp ground white pepper
1 tsp ginger juice
1 tsp sesame oil

⊙⊙⊙ 做法 | Method

1. 石斑肉洗淨，切大件，抹乾水分，用醃料醃 20 分鐘。
2. 燒滾油，將石斑肉撲上生粉，放油中炸至金黃色，瀝乾油分，上碟。
3. 再燒熱鑊，轉慢火，加入牛油和芝士片，煮熔後加入忌廉雞湯，煮滾，淋上石斑肉即成。

1. Rinse grouper and cut into large pieces. Wipe dry and marinate for 20 minutes.
2. Bring oil to the boil. Coat grouper with caltrop starch and deep-fry in oil until golden brown. Drain.
3. Heat a little oil in a wok. Turn to low heat and put in butter and cheese slices. Cook until molten and add cream of chicken soup. Bring to the boil and pour over the grouper. Serve.

入廚貼士 | Cooking Tips

除了加芝士外，也可在石斑肉上灑上芝士粉。

石灣魚腐

Deep-fried Dace Paste

 材料 | Ingredients

鯪魚肉 70 克

70g dace paste

醃料 | Marinade

雞蛋 2 隻
生粉 3 湯匙
油 2 湯匙
鹽 1/2 茶匙
水 1/2 杯

2 eggs
3 tbsps caltrop starch
2 tbsps oil
1/2 tsp salt
1/2 cup water

做法 | Method

1. 鯪魚肉剁幼，加醃料順一個方向攪至起膠，水分多次加入，將魚膠攪成稀糊狀。

2. 燒熱油鑊，用一個已搽油的湯匙，盛起魚膠放入油中，炸至金黃色，撈起上碟，瀝乾油分，可蘸辣椒醬或海鮮醬吃。

1. Chop dace paste until fine. Add in marinade and stir in one direction until sticky. Put in water by several times and stir into a dilute paste.

2. Heat oil in a wok. Scoop the fish paste with a greased tablespoon and put into the oil. Deep-fry until golden brown and drain. Serve with chili sauce or seafood sauce.

入廚貼士 | Cooking Tips

• 鯪魚肉因加入了雞蛋，水分不少，所以要攪拌約 20 分鐘才成稀糊狀，若減去水分效果則不理想。
• 魚腐又可以加入紹菜同煮，味道鮮美。
• 魚腐是石灣的名菜，炸成後似豆腐卜，又滑又香。

煙肉魚棧

Deep-fried Bacon Rolls with Grouper and Dried Black Mushrooms

材料 | Ingredients

石斑肉 400 克	400g grouper fillet
大冬菇 8 朵	8 large dried black mushrooms
煙肉 8 片	8 slices bacon
京葱 2 棵	2 stalks leek

魚肉醃料 | Marinade for fish

糖 1/3 茶匙
鹽 1/8 茶匙
胡椒粉 1/8 茶匙
生粉 1 湯匙（後下）

1/3 tsp sugar
1/8 tsp salt
1/8 tsp ground white pepper
1 tbsp caltrop starch (added at last)

4 人
Serves 4

30 分鐘
30 Minutes

冬菇醃料 |
Marinade for dried black mushrooms

糖 1/2 茶匙
生抽 1/2 茶匙
油 1/4 茶匙

1/2 tsp sugar
1/2 tsp light soy sauce
1/4 tsp oil

漿料 | Paste

雞蛋 2 隻（蛋黃蛋白分開）
粟粉 1/2 杯
麵粉 1/2 杯
水 3/4 杯

2 eggs (separate the egg whites and egg yolks)
1/2 cup corn starch
1/2 cup flour
3/4 cup water

做法 | Method

1. 石斑肉洗淨，切厚方塊，以醃料略醃，備用。
2. 冬菇洗淨浸軟，去蒂，切粗條，以醃料略醃。
3. 煙肉分切兩段，京葱切度。
4. 粟粉和麵粉拌勻，加入蛋黃，慢慢注入清水，調勻成漿料，待發。
5. 蛋白打至企身，加入漿料內拌勻。
6. 冬菇、京葱、石斑肉用煙肉捲實，以牙籤穩固。
7. 燒熱鑊，將煙肉紮泡油，撈起，瀝乾油分，再醮漿料，回鑊炸至金黃色，再瀝乾油分便可。

入廚貼士 | Cooking Tips
魚紮材料可加火腿或西芹。

1. Rinse grouper fillet and cut into thick cubic pieces. Marinate briefly and set aside.
2. Rinse dried black mushrooms and soak until soft. Remove the stalks and cut into thick strips. Marinate briefly.
3. Cut each slice of bacon into 2 sections. Section leek.
4. Mix corn starch and flour. Add in egg yolks. Pour in water slowly. Mix into the paste and set aside for fermentation.
5. Whisk egg whites until stiff and put into the paste. Mix well.
6. Wrap dried black mushrooms, leek and grouper with bacon firmly and fix the ends with toothpicks.
7. Heat oil in a wok. Jiggle the bacon rolls slowly in oil and drain. Coat the rolls with the paste and deep-fry until golden brown. Drain again and Serve.

泰式魚頭煲

Thai Fish Head Pot

⬤⬤⬤ 材料 | Ingredients

三文魚頭 1 個	1 salmon fish head
豬肉 50 克	50g pork
中國芹菜 2 棵	2 stalks Chinese celery
香茅 2 支	2 sticks lemongrass
紅葱頭 2 粒	2 shallots
蒜頭 2 粒	2 cloves garlic

⬤⬤⬤ 魚頭醃料 | Marinade for fish head

酒 1 茶匙	1 tsp wine
胡椒粉 1 茶匙	1 tsp ground white pepper
生粉 1 茶匙	1 tsp caltrop starch
鹽 1/2 茶匙	1/2 tsp salt

Stew

豬肉醃料 | Marinade for pork

鹽、生粉適量

Pinch of salt and caltrop starch

調味料 | Seasonings

青檸汁 1/2 個量
魚露 1 茶匙

1/2 lime (juiced)
1 tsp fish sauce

汁料 | Sauce

雞湯 300 毫升
酸辣香醬 2 湯匙
花生碎 2 湯匙
花生醬 2 湯匙

300ml chicken broth
2 tbsps sour and spicy sauce
2 tbsps chopped peanuts
2 tbsps peanut butter

> **入廚貼士 | Cooking Tips**
> 此菜式可加入其他香料如百里香或九層塔。

做法 | Method

1. 三文魚頭洗淨，抹乾水分，加醃料醃好。
2. 燒熱鑊，三文魚頭放油中炸至金黃色，瀝乾油分備用。
3. 豬肉洗淨，剁碎，加醃料醃好。
4. 紅葱頭、蒜頭、香茅剁茸，中國芹菜切粒。
5. 花生醬用少許水拌勻。
6. 燒熱油鑊，爆香葱茸、蒜茸、香茅及豬肉，放入汁料，加水煮片刻，下魚頭、芹菜和調味料，再煮滾即成。

1. Rinse salmon fish head and wipe dry. Marinate well.
2. Heat oil in a wok. Deep-fry salmon fish head until golden brown and drain.
3. Rinse pork. Chop and marinate.
4. Chop shallots, garlic and lemongrass. Dice Chinese celery.
5. Mix peanut butter with a little water.
6. Heat a little oil in a wok. Stir-fry chopped shallots, chopped garlic, lemongrass and pork until fragrant. Put in the sauce ingredients. Add water and cook for a while. Add fish head, Chinese celery and seasoning. Bring to the boil again and serve.

花生魚頭煲

Peanuts and Fish Head Pot

大魚頭 1 個
花生 100 克
紅棗 10 粒
薑 5 片
水 5 1/2 杯
鹽適量

1 grass carp head
100g peanuts
10 red dates
5 slices ginger
5 1/2 cups water
Pinch of salt

4 人
Serves 4

1 小時 30 分鐘
1 1/2 Hours

⟨⟨⟩⟩ 醃料 | Marinade

胡椒粉 1 茶匙
酒 1 茶匙
鹽 1/2 茶匙

1 tsp ground white pepper
1 tsp wine
1/2 tsp salt

⟨⟨⟩⟩ 做法 | Method

1. 大魚頭洗淨，切開兩邊，瀝乾水分，用醃料醃好。
2. 花生洗淨，紅棗去核，洗淨。
3. 燒熱油鑊，爆香薑片，將大魚頭煎至金黃色，盛起，瀝乾油分。
4. 將水煲滾，加入花生、紅棗和大魚頭，待花生煲腍，加鹽調味即可。

1. Rinse grass carp head and cut into halves along the length. Drain and marinate.
2. Rinse peanuts. Core red dates and rinse.
3. Heat oil in a wok. Stir-fry ginger slices until fragrant. Fry the grass carp head until golden brown and drain.
4. Bring water to the boil. Put in peanuts, red dates and fish head. Cook until peanuts are soft. Season with salt and serve.

入廚貼士 | Cooking Tips
要選體積較大的大魚頭，大魚頭一定要煎過，湯才會變成奶白色。

香芋魚頭煲

Taro and Fish Head Pot

⊙ 材料 | Ingredients

大魚頭 1 個	1 grass carp head
芋頭 300 克	300g taro
中國芹菜 2 棵	2 stalks Chinese celery
薑 2 片	2 slices ginger
生粉適量	Pinch of caltrop starch
油適量	Some oil

⊙ 醃料 | Marinade

酒 1/2 湯匙	1/2 tbsp wine
老抽 1/2 湯匙	1/2 tbsp dark soy sauce
胡椒粉 1 茶匙	1 tsp ground white pepper
鹽 1/2 茶匙	1/2 tsp salt

⃝⃝⃝ 調味料 | Seasonings

上湯 700 毫升	700ml broth
老抽 2 湯匙	2 tbsps dark soy sauce
糖 3/4 湯匙	3/4 tbsp sugar
雞粉 1 茶匙	1 tsp chicken powder
麻油 1 茶匙	1 tsp sesame oil
胡椒粉 1/2 茶匙	1/2 tsp ground white pepper
鹽 1/3 茶匙	1/3 tsp salt

⃝⃝⃝ 做法 | Method

1. 大魚頭洗淨，切開兩邊，用醃料醃 30 分鐘。
2. 燒熱油鑊，將大魚頭沾上生粉，放油中炸至微黃色，撈起，瀝乾油分。
3. 芋頭去皮，洗淨，切厚件，放油中炸片刻。
4. 芹菜去葉，洗淨，切短段。
5. 將芋頭、薑和調味料放煲內煮滾，以慢火煮約 20 分鐘，加入魚頭和芹菜，再煮約 7 分鐘至魚頭熟透，加油適量拌勻，即可。

1. Rinse grass carp head and cut into halves along the length. Marinate for 30 minutes.
2. Heat oil in a wok. Coat fish head with caltrop starch and deep-fry in oil until slightly golden. Drain.
3. Skin taro. Rinse and cut into thick pieces. Deep-fry in oil for a while.
4. Remove leaves from Chinese celery. Rinse and cut into short sections.
5. Put taro, ginger and seasoning into a pot and bring to the boil. Cook over low heat for about 20 minutes. Add fish head and Chinese celery. Cook for about 7 minutes until the fish head is done. Mix in oil and serve.

入廚貼士 | Cooking Tips
魚頭要保持完整，避免弄散，以免魚骨沾着芋頭，令人容易哽骨。

蒜子火腩大鱔煲

Eel and Pork Belly Pot with Garlic

材料 | Ingredients

白鱔 500 克	蒜茸 3 粒量
蒜頭 120 克	葱 2 棵
火腩 80 克	芫茜 2 棵
冬菇 80 克（浸軟）	陳皮 1 片（浸軟）
薑 3 片	酒、生粉適量

500g white eel
120g garlic
80g barbequed pork belly
80g dried black mushrooms (soaked until soft)
3 slices ginger
3 cloves garlic (grated)
2 stalks spring onion
2 stalks coriander
1 quarter dried tangerine peel (soaked until soft)
Some wine and caltrop starch

6 人
Serves 6

30 分鐘
30 Minutes

Stew

汁料 | Sauce

紹酒 1 湯匙　　生抽 2 茶匙
老抽 1 湯匙　　柱侯醬 1 1/2 茶匙
蠔油 1 湯匙　　胡椒粉 1/4 茶匙
糖 2 茶匙　　　水約 90 毫升

1 tbsp Shaoxing wine
1 tbsp dark soy sauce
1 tbsp oyster sauce
2 tsps sugar
2 tsps light soy sauce
1 1/2 tsps Chu Hou sauce
1/4 tsp ground white pepper
Approx. 90ml water

荚汁 | Thickening

生粉、水適量
Some caltrop starch and water

入廚貼士 | Cooking Tips

購買白鱔時，可請魚販用熱水和鹽清洗鱔上的潺。

做法 | Method

1. 白鱔洗淨，汆水以去黏液，切厚片（約 1 1/2 厘米）。
2. 蒜頭洗淨，去衣，燒熱油鑊，下蒜頭炸至金黃色，瀝乾油分備用。
3. 白鱔拍上生粉後，再燒熱油鑊，下白鱔炸至金黃色，瀝乾油分備用。
4. 再燒熱油鑊，爆香薑、蒜茸、陳皮，加入冬菇、火腩爆香，灒酒，再加入汁料，放下白鱔片、蒜頭煮滾後，轉砂鍋內燜 15 分鐘，用生粉水勾荚，再加蔥和芫荽即可。

1. Rinse white eel. Scald for a while to remove the slime. Cut into slices of about 1 1/2cm thick.
2. Rinse garlic and skin. Heat oil in a wok and deep-fry the garlic until golden brown. Drain.
3. Coat white eel with caltrop starch. Heat oil in wok and deep-fry the eel until golden brown. Drain.
4. Heat a little oil in a wok. Stir-fry ginger, grated garlic and dried tangerine peel until fragrant. Add dried black mushrooms and pork belly. Stir-fry until fragrant. Sizzle in wine and pour in the sauce. Put in white eel slices and garlic. Bring to the boil and transfer to a clay pot. Simmer for 15 minutes. Thicken the sauce with caltrop starch solution. Add spring onion and coriander. Serve.

客家豆腐煲

Beancurd Pot in Hakka Style

⊂⊃⊃ 材料 | Ingredients

豆腐卜 200 克	200g beancurd puffs
黃芽白 250 克	250g Tianjian cabbages
鯪魚肉 150 克	150g dace paste
枚頭肉 80 克	80g pork loin
梅香鹹魚 30 克	30g salted fish
蝦米 25 克	25g dried shrimps
芫茜 2 棵	2 stalks coriander
葱 2 條	2 stalks spring onion
薑 1 片	1 slice ginger

醃料 | Marinade

生粉 2 茶匙
胡椒粉 1/2 茶匙
鹽 1/3 茶匙
水 4 湯匙

2 tsps caltrop starch
1/2 tsp ground white pepper
1/3 tsp salt
4 tbsps water

汁料 | Sauce

鹽 1/2 茶匙
水適量

1/2 tsp salt
Some water

入廚貼士 | Cooking Tips
可以用豆腐代替豆腐卜。

做法 | Method

1. 黃芽白洗淨，切短度。
2. 蝦米洗淨，用清水浸軟，瀝乾水分。
3. 鹹魚洗淨，起肉剁幼。
4. 枚頭肉洗淨剁幼，加入鯪魚肉和鹹魚，下醃料，順一個方向攪至起膠，釀入豆腐卜內。
5. 燒熱鑊，下油爆香薑、蝦米，加入黃芽白炒軟，放砂鍋內，再放上豆腐卜，加入汁料煮滾，慢火再煮 20 分鐘，加入芫茜、葱拌勻，原鍋上桌。

1. Rinse Tianjian cabbages and cut into short sections.
2. Rinse dried shrimps. Soak in water until soft and drain.
3. Rinse salted fish. Bone and chop finely.
4. Rinse pork loin and chop finely. Add dace paste and salted fish. Put in the marinade and stir in one direction until sticky. Stuff it into the beancurd puffs.
5. Heat a little oil in a wok. Stir-fry ginger and dried shrimps until fragrant. Add Tianjian cabbages and stir-fry until soft. Put them into a clay pot. Place over beancurd puffs and add the sauce. Bring to the boil. Reduce to low heat and cook for 20 minutes. Mix in coriander and spring onion. Serve.

番茄魚湯

Tomatoes and Fish Tail Soup with Papaya

6 人
Serves 6

1 小時
1 Hour

Stew

◯◯◯ 材料 | Ingredients

大魚尾 1 條	1 grass carp tail
番茄 300 克	300g tomatoes
木瓜 300 克	300g papaya
鮮腐竹 200 克	200g fresh beancurd skin
薑 3 片	3 slices ginger
鹽適量	Pinch of salt

◯◯◯ 做法 | Method

1. 大魚尾洗淨，瀝乾水分。
2. 燒熱油鑊，爆香薑片，下大魚尾煎香，盛起，瀝乾油分。
3. 番茄洗淨，開邊，木瓜洗淨，去皮切大塊。腐竹以濕布抹淨。
4. 燒滾水，放入木瓜、番茄、鮮腐竹、大魚尾同煲約 45 分鐘，下鹽調味即成。

1. Rinse grass carp tail and drain.
2. Heat oil in a wok. Stir-fry ginger slices until fragrant. Fry the grass carp tail until fragrant and drain.
3. Rinse tomatoes and cut into halves along the length. Rinse papaya. Skin and cut into large pieces. Wipe beancurd skin with a wet towel.
4. Bring water to the boil. Add papaya, tomatoes, fresh beancurd skin and grass carp tail. Cook for about 45 minutes and season with salt. Serve.

入廚貼士 | Cooking Tips
番茄不用去皮，多加一些番茄，可令湯的顏色更鮮紅，味道更鮮甜。

南瓜魚湯

Pumpkin and Fish Tail Soup

材料 | Ingredients

大魚尾 1 條	1 grass carp tail
南瓜 500 克	500g pumpkin
黨參 50 克	50g Dang Shen
黑棗 4 粒	4 black dates
豆腐 2 件	2 cubes beancurd
薑 4 片	4 slices ginger
鹽適量	Pinch of salt

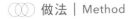

做法 | Method

1. 南瓜洗淨，去皮去瓤，切大塊。黨參、黑棗、豆腐分別洗淨。
2. 大魚尾洗淨，瀝乾水分。
3. 燒熱油鑊，爆香薑片，下大魚尾煎香，盛起，瀝乾油分。
4. 燒滾水，加入南瓜、黨參和黑棗煲 30 分鐘。
5. 加入大魚尾、豆腐同煲 20 分鐘，下鹽調味即成。

1. Rinse pumpkin. Skin and remove the pith. Cut it into large pieces. Rinse Dang Shen, black dates and beancurd respectively.
2. Rinse grass carp tail and drain.
3. Heat oil in a wok. Stir-fry ginger slices until fragrant. Fry the grass carp tail until fragrant and drain.
4. Bring water to the boil. Put in pumpkin, Dang Shen and black dates. Cook for 30 minutes.
5. Add grass carp tail and beancurd. Cook for 20 minutes and season with salt. Serve.

入廚貼士 | Cooking Tips
- 此湯營養豐富、補而不燥。
- 大魚尾可以用大魚頭或鯇魚尾代替。

鹽焗烏頭魚
Salt Baked Mullet

(◯◯) 材料 | Ingredients

....

烏頭魚 400 克
粗鹽 2,000 克
花椒 3 茶匙
八角 3 粒
油 1/2 杯

400g mullet
2kg rock salt
3 tsps Sichuan peppercorns
3 stars aniseeds
1/2 cup oil

4 人
Serves 4

1 小時
1 Hour

Bake

醃料 | Marinade

青檸汁 2 茶匙
胡椒粉 1 茶匙
酒 1 茶匙
鹽少許

2 tsps lime juice
1 tsp ground white pepper
1 tsp wine
Pinch of salt

工具 | Tools

錫紙 1 張
食用沙紙 1 張

1 sheet aluminum foil
1 sheet sandpaper for cooking

做法 | Method

1. 烏頭魚洗淨，抹乾，加入醃料塗勻。
2. 沙紙掃上油，將整條魚包好。
3. 白鑊內加粗鹽、八角及花椒同炒至熱透，再加入油 1/2 杯，炒勻後轉放錫紙上。
4. 把魚放入炒好的鹽中包好，放入預熱的焗爐中以 150℃焗 40 分鐘，取出魚，撕去沙紙，即可上碟。

1. Rinse mullet and wipe dry. Rub over the marinade.
2. Grease sandpaper and wrap the whole fish.
3. Put rock salt, Sichuan peppercorns and aniseeds into a wok without adding oil. Stir-fry until thoroughly hot. Put in 1/2 cup of oil. Stir-fry well and transfer onto the aluminum foil.
4. Put the fish into the salt and wrap well. Bake in a preheated oven at 150°C for 40 minutes. Remove the foil and sandpaper. Serve.

入廚貼士 | Cooking Tips

鹽焗的材料可隨個人喜好而改變。

71

鹽焗秋刀魚

Baked Pacific Saury Fish with Salt

材料 | Ingredients

秋刀魚 3 條
檸檬 1 個
生粉適量

3 pacific saury fish
1 lemon
Pinch of caltrop starch

醃料 | Marinade

鹽 1 1/2 茶匙
胡椒粉 1 茶匙
酒 1 茶匙

1 1/2 tsps salt
1 tsp ground white pepper
1 tsp wine

做法 | Method

1. 秋刀魚去除腸臟，洗淨，抹乾水分，用醃料醃好，魚面塗抹上生粉。
2. 秋刀魚放入已預熱的爐內，以 180℃焗至金黃色，將魚反轉，再焗至金黃色即可。
3. 進食時灑上檸檬汁即可。

1. Remove entrails from the fish. Rinse and wipe dry. Marinate well and rub caltrop starch over the surface.
2. Bake the fish in a preheated oven at 180℃ until golden brown. Turn the fish to the other side and bake until golden brown.
3. Sprinkle over lemon juice and serve.

入廚貼士 | Cooking Tips
不用焗爐，可改用鑊煎至兩面香脆。

涼瓜魚茸焗蛋白

Baked Egg Whites with Bitter
Melon and Dace

材料 | Ingredients

涼瓜 1 個　　　　葱粒適量
鯪魚肉 80 克　　鹽適量
蛋白 8 隻　　　　生粉適量
陳皮碎 1 片量

1 bitter melon
80g dace paste
8 egg whites
1 quarter dried tangerine peel (chopped)
Some diced spring onion
Pinch of salt
Pinch of caltrop starch

4 人
Serves 4

30 分鐘
30 Minutes

調味料 | Seasonings

生粉 1 茶匙	1 tsp caltrop starch
胡椒粉 1/4 茶匙	1/4 tsp ground white pepper
糖 1/4 茶匙	1/4 tsp sugar
鹽 1/8 茶匙	1/8 tsp salt
水約 250 毫升	Approx. 250ml water

做法 | Method

1. 涼瓜洗淨，瀝乾水分，開邊，去瓤切薄片，用鹽搓勻，待片刻，洗淨瀝乾水分。
2. 鯪魚肉加少許鹽、生粉，順一個方向攪至起膠。
3. 蛋白打勻，加入調味料，逐少加入鯪魚肉和涼瓜片、陳皮碎、葱粒和少許油拌勻。
4. 將蛋白混合物倒在淺的焗盆中，放入已預熱焗爐用 200℃焗 20 分鐘即可。

1. Rinse bitter melon and drain. Cut into halves along the length. Remove the pith and cut into thin slices. Rub over salt and set aside for a while. Rinse and drain.
2. Add a little salt and caltrop starch into dace paste. Stir in one direction until sticky.
3. Whisk egg whites. Mix in seasoning. Put in dace paste, bitter melon slices, chopped dried tangerine peel, diced spring onion and a little oil bit by bit. Mix well.
4. Pour egg white mixture into a shallow baking tray. Bake in a preheated oven at 200°C for 20 minutes. Serve.

入廚貼士 | Cooking Tips
與其他海鮮一樣，鯪魚肉要順一個方向攪至起膠，否則會鬆散。

吞拿魚撻

Baked Tuna Tart

◯◯ 材料 | Ingredients

酥皮
麵粉 170 克
牛油 50 克
蛋黃 1 隻
鹽 1/2 茶匙
水 50 毫升

Pastry
170g flour
50g butter
1 egg yolk
1/2 tsp salt
50ml water

餡料 | Fillings

罐裝吞拿魚 100 克（水浸）
粟米粒 200 克
火腿 2 片（約 1 厘米厚）
沙律醬 100 克
芝士粉適量

100g canned tuna in water
200g corn kernels
2 slices ham (about 1 cm thick)
100g salad dressing
Pinch of cheese powder

調味料 | Seasonings

鹽 1/2 茶匙
胡椒粉 1/2 茶匙

1/2 tsp salt
1/2 tsp ground white pepper

入廚貼士 | Cooking Tips

- 麵粉篩過才不會起粒。
- 火腿如改用煙肉便不用加鹽。

做法 | Method

1. 麵粉篩好，牛油切碎，用手指搓碎成麵包糠狀，加入蛋黃和鹽拌勻，慢慢地加水，搓成幼滑粉糰，待發 30 分鐘。
2. 用木棍擀薄粉糰，用圓模吸出多個小圓塊，鋪入已塗油的撻模內，用針刺多個小孔。
3. 放入已預熱的焗爐內，以 200℃焗約 10 分鐘，待涼。
4. 火腿洗淨，切粒，粟米和吞拿魚瀝乾水分，加入調味料和沙律醬拌勻，放入焗好的撻皮內，灑上芝士粉即成。

1. Sieve flour and cut butter into chips. Mix them and knead with fingers into breadcrumbs like. Mix in egg yolk and salt. Add in water slowly and knead into a fine dough. Set aside for 30 minutes.
2. Roll flat the dough with a wooden roller. Mould out many small round pieces with a round mould. Line the dough into greased tart moulds. Pierce many small holes with a pin.
3. Bake in a preheated oven at 200℃ for about 10 minutes. Set aside to let cool.
4. Rinse ham and dice. Drain corns and tuna. Mix in seasoning and salad dressing. Put the mixture onto the tarts. Sprinkle over cheese powder and serve.

吞拿魚薯茸批

Tuan and Potato Mash Pie

材料 | Ingredients

罐裝吞拿魚 100 克（水浸）
蘑菇 100 克
薯仔 3 個
洋葱 1/2 個
蒜頭 2 粒
牛油 1 湯匙
淡奶 2 湯匙
粟粉 2 湯匙
鹽 3/4 茶匙

100g canned tuna in water
100g button mushrooms
3 potatoes
1/2 onion
2 cloves garlic
1 tbsp butter
2 tbsps evaporated milk
2 tbsps corn starch
3/4 tsp salt

6 人
Serves 6

45 分鐘
45 Minutes

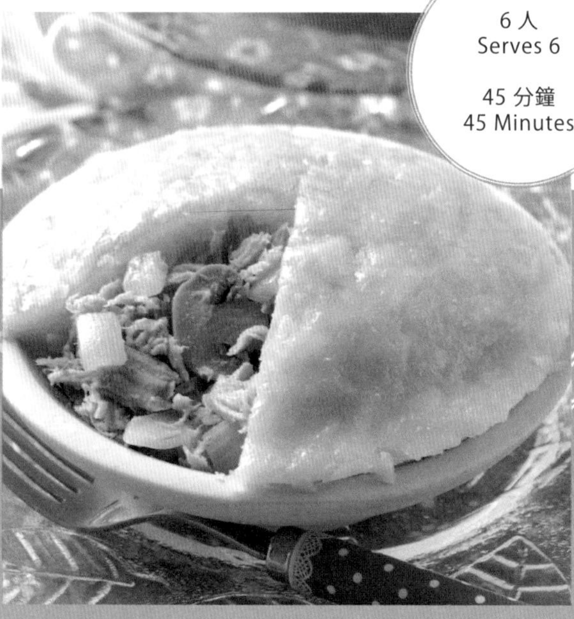

Bake

⊗⊗ 做法 | Method

1. 薯仔洗淨，去皮切片，隔水蒸腍，壓成茸，加入牛油和鹽拌勻。

2. 吞拿魚瀝乾水分。粟粉加淡奶拌勻成淡奶漿。

3. 洋葱洗淨，去衣切粒，蒜頭洗淨，去衣剁茸，蘑菇洗淨，切片。

4. 燒熱鑊，爆香洋葱粒、蒜茸，加入吞拿魚和蘑菇片，倒下淡奶漿炒勻，盛起。

5. 將一半薯茸平鋪在圓模上，將炒好的材料放薯茸上，最後鋪上餘下的薯茸，放入已預熱的焗爐，以 180℃焗約 20 分鐘即成。

1. Rinse potatoes. Peel and slice. Steam until soft and crush well. Mix in butter and salt.

2. Drain tuna. Mix corn starch and evaporated milk to make the evaporated milk paste.

3. Rinse onion. Skin and dice. Rinse garlic. Skin and chop. Rinse button mushrooms and slice.

4. Heat a little oil in a wok. Stir-fry diced onion and chopped garlic until fragrant. Put in tuna and button mushrooms slices. Pour in the evaporated milk paste. Stir-fry well.

5. Spread half amount of potato mash over the round mould. Put stir-fried ingredients on top and lastly spread the remaining potato mash. Bake in a preheated oven at 180°C for about 20 minutes. Serve.

入廚貼士 | Cooking Tips

可用新鮮的吞拿魚代替罐裝吞拿魚。

古法蒸桂花魚

Steamed Mandarin Fish in Traditional Style

⬤⬤ 材料 | Ingredients

桂花魚 1 條	1 mandarin fish
瘦肉 40 克	40g lean pork
冬菇 2 小朵	2 small dried black mushrooms
葱 2 棵	2 stalks spring onion
薑 2 片	2 slices ginger

⬤⬤ 魚醃料 | Marinade for fish

鹽 1/2 茶匙	1/2 tsp salt
酒 1/2 茶匙	1/2 tsp wine
胡椒粉 1/2 茶匙	1/2 tsp ground white pepper

⊙⊙ 肉醃料 | Marinade for meat

生抽 1/4 茶匙
生粉 1/4 茶匙

1/4 tsp light soy sauce
1/4 tsp caltrop starch

入廚貼士 | Cooking Tips

蒸魚的汁比較腥，所以要倒
去，再煮豉油淋在魚面上。

⊙⊙ 調味料 | Seasonings

生抽 1 湯匙
老抽 1 茶匙
糖 1/2 茶匙
水 1 湯匙

1 tbsp light soy sauce
1 tsp dark soy sauce
1/2 tsp sugar
1 tbsp water

⊙⊙ 做法 | Method

1. 桂花魚去鱗、去內臟，洗淨，抹乾水分，加醃料醃好，放碟上。
2. 瘦肉洗淨，切絲，加醃料醃好。
3. 冬菇浸軟，去蒂切絲，薑、葱洗淨切絲。
4. 在桂花魚上放瘦肉絲、冬菇絲和一半薑絲，蒸 10 分鐘，倒去蒸魚的汁。
5. 燒熱鑊，爆香餘下的薑絲，加入調味料煮滾，在桂花魚上放下葱絲，淋上煮滾的汁料即成。

1. Scale and remove entrails from mandarin fish. Rinse and wipe dry. Marinate well and put onto a plate.
2. Rinse lean pork and cut into shreds. Marinate well.
3. Soak dried black mushrooms until soft. Remove stalks and cut into shreds. Rinse ginger and spring onion and cut into shreds.
4. Put shredded pork, shredded dried black mushrooms and half amount of ginger shreds above the mandarin fish. Steam for 10 minutes and remove the extract.
5. Heat a little oil in a wok. Stir-fry the remaining ginger shreds until fragrant. Put in seasoning and bring to the boil. Put spring onion shreds on top of the mandarin fish and pour over boiled seasoning sauce. Serve.

紫翠香魚

Steamed Nori Seaweed Fish Rolls

⟨◯◯◯⟩ 材料 | Ingredients

鯪魚肉 200 克
紫菜 4 片（方型）
生菜絲少許

200g dace paste
4 slices square-shaped nori seaweed
Some lettuce shreds

4 人
Serves 4

20 分鐘
20 Minutes

醃料 | Marinade

鹽 1 茶匙
胡椒粉 1 茶匙
麻油 1 茶匙

1 tsp salt
1 tsp ground white pepper
1 tsp sesame oil

芡汁 | Thickening

生粉 2 茶匙
雞粉 2 茶匙
水 120 毫升

2 tsps caltrop starch
2 tsps chicken powder
120ml water

做法 | Method

1. 魚肉加醃料拌勻，順一個方向攪至起膠。
2. 取一張紫菜平放碟上，將魚膠均勻地塗在紫菜上，向內捲成長條形，隔水以大火蒸 5 分鐘，取出，待涼後切成小片。
3. 生菜絲放在碟上，放上魚卷。
4. 燒熱鑊，煮滾芡汁，淋在魚卷上即成。

1. Mix dace paste with the marinade. Stir in one direction until sticky.
2. Put a slice of nori seaweed onto a plate. Spread dace paste evenly onto the nori seaweed. Roll up into long strips. Steam over high heat for 5 minutes. Set aside to let cool and cut into small slices.
3. Place lettuce shreds onto a plate and put the fish rolls on top.
4. Heat oil in a wok. Add the sauce and bring to the boil. Pour over the fish rolls and serve.

入廚貴 | Cooking Tips

- 紫菜可改用蛋皮，即用雞蛋煎成薄餅皮，紫翠香魚即成黃金香魚。
- 蒸好的魚卷切片後，可放鑊中煎香。

鬼馬蒸魚腸

Steamed Egg and Fish Intestines with Deep-fried Breadsticks

材料 | Ingredients

鯇魚腸 2 份
雞蛋 3 隻
油條 1 條

2 portions grass carp intestines
3 eggs
1 deep-fried fluffy dough stick

....

胡椒粉 1 茶匙
鹽 1/2 茶匙

1 tsp ground white pepper
1/2 tsp salt

◯◯◯ 做法 | Method
....

1. 油條切 1 厘米厚片。

2. 鯇魚腸剪開，去除脂肪，洗淨，加入醃料醃好，保留水分。

3. 雞蛋打勻，加入魚腸，放鑊中蒸約 10 分鐘，將油條平鋪在將熟
   的蛋面上，再蒸 5 分鐘即可。

1. Cut dough stick into slices of 1cm thick.

2. Cut open grass carp intestines. Remove fat and rinse. Marinate
   and reserve water.

3. Whisk eggs and add in fish intestines. Steam for about 10
   minutes. Spread dough stick slices over almost cooked steamed
   eggs. Steam for 5 minutes more and serve.

入廚貼士 | Cooking Tips

用粗鹽擦洗魚腸，去除不潔物和脂肪時，宜先除去魚膽，否則不小
心弄破會令魚腸味道變苦。

Stewed Dried Fish Maw
and Chicken Soup

魚肚燉雞湯

6 人
Serves 6

5 小時
5 Hours

⦿ 材料 | Ingredients

老雞 1 隻   1 mature chicken
魚肚 80 克   80g dried fish maw
金華火腿 40 克 40g Jinhua ham
杞子 20 克   20g Qi Zi
生薑 3 片   3 slices raw ginger
鹽適量   Pinch of salt

⦿ 做法 | Method

1. 老雞洗淨，去皮，汆水，洗淨。

2. 魚肚洗淨後，用清水略為浸泡。其餘材料洗淨。

3. 將所有材料同放一個大燉盅內，加入凍開水，隔水燉 4~5 小時，
   加鹽調味即可。

1. Rinse mature chicken. Skin and scald for a while. Rinse.
2. Rinse dried fish maw and soak in water briefly. Rinse the
   remaining ingredients.
3. Put all ingredients into a large stewing pot. Add cold drinking
   water and stew for 4~5 hours. Season with salt and serve.

入廚貼士 | Cooking Tips
魚肚是滋陰的食品，而且補而不燥。

編著
梁燕

編輯
Pheona Tse　Kitty Chan

美術設計
Venus

排版
劉葉青

翻譯
梁悅冰

攝影
家家

出版者
萬里機構出版有限公司
香港鰂魚涌英皇道1065號東達中心1305室
電話：2564 7511
傳真：2565 5539
電郵：info@wanlibk.com
網址：http://www.wanlibk.com
　　　http://www.facebook.com/wanlibk

發行者
香港聯合書刊物流有限公司
香港新界大埔汀麗路36號
中華商務印刷大廈3字樓
電話：2150 2100
傳真：2407 3062
電郵：info@suplogistics.com.hk

承印者
中華商務彩色印刷有限公司
香港新界大埔汀麗路 36 號

出版日期
二零一九年三月第一次印刷

萬里機構

萬里 Facebook